HOW GOOGLE TESTS SOFTWARE

Peter Wright

Copyright © 2018

All Rights Reserved, Peter Wright

The author reserves all the rights to this book. The author does not permit anyone to reproduce or transmit any part of this book through any means or form be it electronic or mechanical. No one has the right to store the information herein in a retrieval system, or to photocopy, record copies, scan parts, etc., without the proper permission of the publisher or author.

Disclaimer

All the information in this document is to be used for informational and educational purposes only. The author will not account in any way for any results that stem from the use of the information herein. While conscious and creative attempts have been made to ensure that all information provided herein is as accurate and useful as possible, the author is not legally bound to be responsible for any damage caused by the accuracy as well as use/misuse of this information.

Contents

Preface 4

Introduction 5

Quality Testing 6

Combining Testing and Development 8

Internal Employee Testing 11

Crowd-Testing Platforms 14

Dogfooding 16

Beta Testing 18

Code Coverage 20

Conclusion 23

Preface

As one of the leading technology companies in the world, Google produces a ton of software. From Web-based products like Google Search and Google Translate to Desktop Applications like Google Chrome and Google Drive, software plays a very crucial role in Google's existence as a company. As a result, Google pays a lot of attention to the quality of software it produces. Considering the sheer amount of software built by Google, however, one question that is often asked from engineers and employees at Google is "How does Google test software?" How does a company as large as company undergo the testing of its various software products to make sure that every software program and application released out to the public is of the best quality and standard?

This book will offer readers insight into the Google software testing process, including the various stages of the process, the aspects Google considers to be essential, and what exactly software quality means to Google.

Introduction

Testing plays a crucial part in the development of software and programs. It is essential because it helps to point out any possible errors and defects software might have early on while it is still in the development phase. And with some Google software having as much as two billion lines of code, it goes without saying that testing as to be carried out on Google software right from early on in the development stage to avoid releasing a faulty program to its customers.

Without having high standards for software development, Google would not be the company it is today, producing software for a diverse range of platforms, devices, and functions, the Google software testing process signifies Google's drive as a company to continually provide quality products that serve the purposes they were created to carry out.

Quality Testing

As far as Google is concerned, passing a test does not equate to a product being a quality one. Quality is inherent in products, and it is not tested in. If the quality of a program is to be proven, it means it has to have started from when it was being written. However, while quality cannot be tested into software or program, it is also impossible to develop quality software without testing it. There is no way to guarantee that a product is as high of a quality as the developer says it is without testing it.

As such, Google's approach to ensuring the quality of its software is to stop separating software development and software testing and instead integrate both processes. For Google, testing isn't a different aspect of the product development process. It is an integral part of it. At Google, quality is achieved by combining development and testing until one process cannot be separated from the other. This combination not only helps boosts the developer's confidence in the software he is developing, but it also helps speed up the process of testing newly developed software, thus reducing the time-to-market for this software.

Combining Testing and Development

As one of the foremost technology-based companies in the world, Google pays a lot of attention to the software it produces. To ensure this software are top notch, it adopts a rigorous approach to their development of this software.

Unlike most tech companies on an almost equal footing with Google, Google does not have a separate testing department in charge of overseeing the testing process of the different software Google produces. Instead, the test aspect of Google can be found in a Focus Area known as Engineering Productivity, and this Focus Area is largely responsible for Google's testing process. It comprises:

i. A product team made up of different types of engineers from the various fields Google employs from. This team is responsible for building and maintaining all the tools, analysers, and systems used in the software development process at Google.

ii. A services team which offers expertise to product teams at Google on several relevant topics like testing, documentation, and using tools.

iii. Embedded engineers who assist Google product teams when they are needed. Some of these engineers might be assigned to a team for extended periods of time while others move through teams when they are needed.

However, while Engineering productivity members are there to assist the product teams in accomplishing their tasks, the responsibility of maintaining the quality of software programs generated still rests mainly on the product teams, instead of the testers. Every developer is expected to test his code and software properly, and the job of the tester is to make sure the developers have access to the automation infrastructure that aid testing, as well as the enabling processes that contribute to this self-reliance. Testers are essentially present there to allow developers to be able to test the software themselves. This allows the developers to test the program as they write it.

The advantage inherent in combining development and testing is that engineers and developers can test the products they create as each stage is accomplished, allowing them to ensure that the software base is solid, in addition to helping to avoid building up the software on a foundation that will eventually come crashing down. Also, with over two billion lines of code, it is better to have the engineers test the code and spot any potential bugs early on, so any issues that may occur because of these errors are taken care of while the software is still in development.

Internal Employee Testing

The first step in Google's software testing process is the Internal Employee Testing. As one of the world's foremost search engine companies, maintaining the functionality of Google Search is very critical to Google's existence as a company. As such, the team charged with maintaining the Google Search framework also operates large and rigorous testing machinery. Google starts its testing procedure by releasing its software to be tested by its wide array of dedicated internal employees.

As a company, Google itself has a disproportionately low number of testers compared to other tech companies of a similar level. Because of this fact, a lot of the testing requirements for their products is met by the developer. The idea behind having the developers pay a lot of focus to testing while also developing the apps is because the best person to test something is often the person who built it. As the developer, they have an idea of how many lines of code went into developing the program, as well as knowing what these lines of code are supposed to do. As the program authors, the developers are usually the best people to find bugs in the program.

The Engineering Productivity Focus Area of Google plays a vital part in this part of testing because they provide and maintain the tools that allow engineers and developers test the software they manufacture. Use of this software is what allows Google engineers and developers to test the software they write. Embedded engineers who are part of the Engineering Productivity team also contribute as they have experience in a variety of processes and software, and thus help with crosschecking and proofing the work of the Product teams, thus helping to guide the projects from conception to when they are launched.

The advantage to using this method is that it works more towards prevention than towards detection. By having the developer run through that program and checking for errors in the code, there is less chance of mistakes or errors being committed. By embedding the practice of regular software testing in the software development process, the process is quickened as there is less chance of bugs getting through this phase and entering other phases, thus cancelling out the need to frequently return the software to the programmers when errors are spotted, which would necessitate restarting the whole testing process again. This also greatly reduces the number of dedicated testers Google needs to employ at a time. It also incentivises the developers to work more towards error and bug prevention than having them rely on the use of dedicated testers to spot whatever bugs may exist in the programs they write.

Crowd-Testing Platforms

Another important aspect of the Google software testing machinery is the use of crowd testing. The main reason this plays a part in the Google testing process is that using Crowdsourced platform offers Google the opportunity to have its apps tested in real-world conditions. Crowd testing, as the name implies, involves using crowds to test software under real-world conditions. It allows for more individuals to participate, usually at a reduced cost and with better testing coverage. In using Crowd Testing, Google releases copies of software under development to crowd testing sites, which have crowd testers make use of these products and provide Google with feedback on the performance of the products.

Crowd testing allows Google to align and synchronise its testing efforts with their plans and goals for software development. By adopting crowd testing platforms as part of their testing process, Google can get insights into the performance of their software in different test environments. This helps to spot a variety of errors and bugs faster, allowing the development team to resolve these problems as they present themselves and improve the program capabilities. With the insights generated when real life users make use of these programs, Google is also able to make its programs and applications better as each build and update is released.

Crowd testing also offers Google insight into how the programs being tested perform out of the developer's lab. It gives insight into the program's performance in the real world, allowing Google to figure out critical issues that need to be fixed immediately. This removes the risk of the software failing later on after being deployed on a larger scale. This further helps reduce the software's time to market as it eliminates the need for longer timeframes that traditional testing methods would require to generate insights.

Dogfooding

Dogfooding is a testing practice which involves a company's staff using software the company creates to work out any kinks and spot bugs before the program is rolled out to the general public. It is a common method used at Google to test the quality of their programs by having members of their staff use the software in the way a customer would, helping to spot program glitches and bugs by using them in real life. The advantage to using this method is that by having employees and people in the company test it, you can make sure the program works as it should while also making incremental improvements to the program where necessary.

Dogfooding also provides Google with regular feedback on the quality of programs being churned out and how well they perform on different platforms and when used for various tasks. Also, having the trained personnel at Google use the applications as the customer is expected to also helps with spotting errors faster as their training allows them to figure out what part of the programs are functioning correctly and what parts are misbehaving as a result of bugs and glitches. With dogfooding, feedback can also be easily and quickly communicated back to the developer through internal means of communication.

Software developers and companies who make use of their programs are also able to see first-hand how the program works, as well as being able to spot whatever problems may emerge in the course of using the product. As a side benefit, it helps with marketing the product as it looks better and inspires confidence when you use your products and programs, as opposed to using that of a competitor. It helps demonstrate the capabilities of the program to the market as they can see it being used by the software developer too.

Beta Testing

In Google's product testing process, the beta test is the final stage before the program is released to the general public. It involves distributing the product to individual users who have signed up to be beta testers for the products or beta testing sets where beta versions of products are released to testers. When users sign up to be beta testers, they are notified every time a beta version of a Google program is released. They can then download a copy of the program, often a new program or an already existing one with new features that are just being rolled out.

As beta testers, participants in this stage of testing are often required to sign agreements with Google, not to divulge sensitive information about the program being tested to the general public. Google also releases a set of guidelines detailing instructions which beta testers are required to follow. These include attempting a number of tasks with the program and reporting back any bugs to Google.

Input supplied through beta testing helps enhance the product quality, further boosting the chance of the program's success when it is released out to the public. Beta testing offers Google an overview of the functionality of the programs they are working on at a time. Because it possibly involves numerous users running the program across a multiplicity of different systems, it quickens up the process of ensuring that an application works on and is compatible with at least a majority of systems and platforms it was designed for. Any errors or bugs spotted in this stage can be resolved, enabling the program to be compatible across as many platforms as it was intended for. As the final testing process, Beta testing is also very instrumental in helping to discover any bugs, code errors, and gaps in the program that may have slipped through the other three testing stages discussed above.

Code Coverage

Code coverage also plays an integral part in the Google program testing process. It refers to how much of a program's source code is covered by a testing plan. Code coverage is also known as test coverage. It describes how many lines of code bundled in a testing program is covered in every of the software testing processes listed above. By using code coverage to analyse the testing process, the software developer can ascertain that their all aspects of their software have been tested for possible bugs and kinks and is relatively free of any major or even minor errors. Engineers at Google are allowed to make use of various Integrated Development Environments, based on their preference as well as make use of the code coverage testing tools these various IDEs offer.

Code coverage is stated as a percentage which is derived by calculating the amount of the software's source code which has been executed in the software testing process. Whatever percentage is given as the code coverage amount means that that is the percentage of the software's source code that has been executed during testing. It serves to show how reliable exactly a program is before it is released. At Google, a certain level of code coverage is required at the testing stage before any software is allowed to roll out to customers.

One major benefit of code coverage in the Google software testing process is that it helps to identify any extra lines of code that are not functional or needed for the application to run as it was intended. This is beneficial because having extra code that has no use only increases the risk of the software malfunctioning and removing it helps reduce the software package size. Also, it helps make the code base easier to maintain as the software is extended because it reduces the effort required to test the software in the future. Code coverage analysis also helps increase software quality and boost customer satisfaction. In addition, it results in faster time to market as it reduces the time it would take to test the software. Because developers can detect which parts of the software have been tested, and which part have not, they can direct their testing efforts into those specific parts instead of focusing on the program as a whole.

Conclusion

The core part of Google's software testing process involves four stages, all of which are crucial to Google's ability to eventually come up with quality software. This process starts with internal employee testing, where engineers within Google take part in testing the product. This is then followed by the crowd-testing stage where Google sends this software out to paid testers, to test the software on a larger scale. After this stage, the software is then subjected to Dogfooding, a testing stage where Google employees are required to make use of the software that has been developed, allowing it to be tested on a larger scale. The final part of Google's testing process is the Beta Testing stage, where the finished software is sent out to a small portion of Google clients and users who have signed up to be part of the beta testing phase. These clients and users are to use this software and report back to Google about any bugs or issues they might have encountered in the course of using the product, so the engineers can give the product a final look over before it can then be opened up to the general public for use.

As the idea at Google states, quality cannot be tested into a product. Regardless of the number of tests you run on a product, if it was never a quality product from the start, the tests will not make it one. This is one of the reasons why Google encourages its developers to constantly test their work as they write it in the early phases, even before other stages of testing are embarked upon. By constantly testing every stage of their work as they do it, Google developers can create software of high quality, which is often shown during the testing process. By ensuring that all of its software is well tested, Google further strengthens its position as a foremost tech company in the world.

www.ingramcontent.com/pod-product-compliance
Lightning Source LLC
Chambersburg PA
CBHW031524210526
45464CB00007B/3015